Earth Physics with Recent Advances

About the Author

Prof. J.N. Nanda besides being an Eminent Scientist with two earned doctorates in Physics and Applied Physics respectively has also been active in the social and cultural sphere. He was the first scientist in India to begin research work in oceanography and underwater sound. He prefers the advance of co-existence rather than aiming at unity in diversity. Understanding the complexity of human mind, he advocates a new Mahayana of co-existence and earnest efforts to square beliefs with advances in science and technology. He has numerous papers and books to his name. He has travelled extensively and led some government delegations abroad.

EARTH PHYSICS
WITH
RECENT ADVANCES

J.N. Nanda

CONCEPT PUBLISHING COMPANY PVT. LTD.
NEW DELHI-110059

ISBN-13: 978-93-5125-177-4

First Published 2015

Published and Printed by

Concept
Since 1974

Concept Publishing Company Pvt. Ltd.
A/15-16, Commercial Block, Mohan Garden
New Delhi-110059 (India)
T : 25351460, 25351794, *F* : 091-11-25357109
E: publishing@conceptpub.com, W: www.conceptpub.com
Editorial Office: H-13, Bali Nagar, New Delhi-110 015, India.

Cataloging in Publication Data--
Courtesy: D.K. Agencies (P) Ltd. <docinfo@dkagencies.com>

Nanda, J. N., 1920- author.
Earth physics with recent advances / J.N. Nanda.
pages cm
Includes bibliographical references (pages) and index.
ISBN 9789351251774

1. Geophysics. 2. Magnetism. I. Title.

DDC 550 23

Preface

Earth Physics has advanced considerably in the last few year and so many new factors have been added that most scientists are now become ignorant of modern earth physics. This book is a comprehensive outline of physics of the earth incorporating most of the new advances. It is being prepared by the author who has contributed many of the innovative insights into the subject having been working on earth physics and oceanography issues since 1948 and has discovered that most scientists need some compilation to guide them across the new insights A new theory rarely develops in step by step method. It usually rests on a jump-step discovered by an individual scientist who is deeply involved. The rungs of the ladder reaching to the insights are set up later.

This is how Feynman or Einstein constructed new theories of Physics that have revolutionized human civilization. Leaving aside the philosophy of scientific progress, we have in this book tried to address the common man. We start from basics which are already put together laboriously to form the subject of earth physics but which is smothered by experimental and visual data gathered over centuries.

Reference to original papers is scattered all over the book.

I owe sincere thanks to the staff of the Department of Terrestrial magnetism (USA) and its Director Dr. M.A. Tuve who selected me as a post doctoral Research Fellow in 1948 specifically to bring back to life the subject of earth physics which seemed to have reached still waters. In consultation with him I chose the subject of microseisms which involves a study of sea waves as well as the travel of seismic waves. His department became my alma mater for geophysics and oceanography. I had the good fortune of being the only physical oceanographer in India to participate in a Commonwealth Conference on Oceanography in 1954 in London and ever since then I had retained deep interest in oceanography and published research work in this field as well as helped Andhra and Kerala Universities in India to start teaching oceanography, in close association with Prof. E.C. LaFond of USA on a teaching assignment in India. He made it possible for me to participate in the Downwind Expedition of the Scripps Institute of Oceanography. I carried out my experimental work for finding an instant method of determining relative amounts of biological content in any column of sea–water, a task picked up by me when attending the oceanography meeting at NIO of U.K. I must also add my thanks to Min. of Defence who employed me on return to India in 1951 after advanced training in USA. Dr. M.A. Tuve, the Director of the Department of Terrestrial Magnetism of the Carnegie Institution of Washington had donated some equipment to me for work in oceanography and microseisms started in USA and my job of setting up a naval physical and oceanography laboratory for the Indian Navy was of great help. After my retirement from the Defence appointments in India principally as Director of Defence Materials Research Laboratory at Kanpur and Director and Dean of the Institute of Armament Technology at Pune, I had the post retirement jobs as the Executive Secretary of the Ind. National Science Academy and a Visiting Professorship for Geophysics at the Punjab University at Chandigarh. I am grateful to Dr. A.R. Kidwai for creating an honorary post of Director (Science) in his Zaheer Science Foundation at Delhi with complete freedom to carry on my scientific work.

I should also record my appreciation of considerable help from my son Dinesh Nanda, the scientist in charge of exploration work of Indian Oil and Natural Gas Commission. Thanks are due to Dr. S.C. Solomon, a former director of DTM and an authority on rocket sounding of Mars, who has often cleared my doubts and answered my questions about earth and Planetary Physics.

The advanced earth physics described in this little book will provide the key to many yet unsolved problems and lead to yet unexplored extensions of earth physics. Here, I shall within this preface describe a simple solution to the oft repeated observation from study of ancient rocks that the earth had a heavy bombardment at least on two occasions after an interval of few hundred million years after earth's formation. The textbook earth physics have no explanation but the recent advances explained in this book has a ready answer. There was no bombardment from any other heavenly body but it was the return under gravity of particles blown out of the mass which was to become earth during and just after the formation of the earth. The earth was formed by a series of expulsions in small quantities of earth material from a suitable pocket of the sun. Such material spent some time in orbiting around the earth before falling to the growing earth under its gravity. After about 700 million years of the formation of the earth, the earth had its first atomic explosion which also sent out particles and their return was the second bombardment.

Contents

Introduction

New Factors in Earth Physics

Earth science needs a fresh look at the problems it faces. There are claims that an earth dynamo based on convection currents inside the earth may explain both the magnetic field of the earth and its reversals. Will then there be a different theory for the emergence of a field on the surface of Mercury which has no possibility of convection currents in it. Similarly, there is emergence of helium from inside the earth? This helium escapes into space after a long residence period in the earth's atmosphere. For reversals and repeated establishment of field as well as for many small scale convections of the magma in the mantle, considerable energy input must be there over what comes from sunshine. The estimate of James is that an input of energy of more than 10^{n}W is necessary. The possible growth of the moon so near a gravitating body like the earth was still shrouded in mystery.

Keeping these blocks in the progress of earth physics in mind, Dr. M.A Tuve selected me after my acceptance of a post graduate Carnegie Fellowship in his Department of Terrestrial Magnetism at

Washington D.C. enjoined me this task of removing such hurdles from the progress of earth physics and even assure my continuing such work when I return to India after the completion of training in USA at the expense of Indian post war planning. The DTM also sent along with me some equipment for my research in microseisms and oceanography that I had started in DTM. My preliminary studies had already shown that some new concepts that I had hit upon, if introduced might resolve some difficulties. My theoretical ideas were borne out by my research and in addition they alone could logically explain some observations of other scientists abroad when the experimenters had no logical explanation. This book deals with standard but modern geology as well as new factors introduced in Earth Physics.

The new factors mainly involve three aspects: (a) ^{235}U, a spontaneously fissioning element is present from start in all planets and invariably causes atomic explosions in planets, (b) planetary magnetic fields do not arise from any convection driven planetary exciting magnetic field, (c) the moon has grown out of planetary explosion driven scabs and stones from the earth, (d) I have also discovered that earth has a special position among solar planets, (e) that much more study has to be made of the solar corona that delivers planetary material in driblets to planet sites, (f) I have also something to say about the earth being the only planet ever likely to grow human civilization, and (g) our existence on earth is likely to be disturbed by the travails against ice ages more than by problems from atmospheric or ocean storms or earthquakes than by any violent impacts with other heavenly bodies for the remaining life of earth or of the sun.

The earth's special position in the solar system is discussed in chapter 1. The earth being equipped with special provisions so necessary for formation of living moities depends primarily because the earth is the only planet which has copious availability of nitrogen so necessary in the living cell formation along with other necessary conditions of moderate temperature and flowing water. The usual theory of growth of the high wave that inundates the coastal area when the center of the tropical cyclone or the hurricane crosses the sea shore gives the height of the cyclonic wave at least 3 times too low. The high wave calculated by me for the Andhra cyclone of 1972 gives the actual wave height. The India Meteorological Department was duly impressed but would not go for announcing the new result when the cyclone had already done its damage and the warnings had been given based on the long fetch of the winds. The new theory could usefully be employed in the warning set-up in USA regarding hurricane flooding the shores of the southern states affected by hurricanes in the Mexican waters.

Ice ages have occurred when an ice sheet spreads over most of the earth's surface. The glaciation is supposed to proceed further or dwindle depending on the variation in the solar energy reaching the earth or due to drifting of continents or changes in the straits, connecting the oceans which may disrupt the circulation of sea water. The current stage is the aftermath of the fourth ice age where human civilization has suddenly emerged and blossomed. There is so much agitation because of increase in CO_2 and refrigeration gases in the atmosphere. In fact it is the blanket effect of these gases or ammonia that helped at one stage to keep the atmosphere hot when solar energy received by earth had fallen.

The last new factor is the action of the plasma of the corona on the earth material when it transports the same to the earth against solar gravity. In case the corona happens to reach the earth surface just after a nuclear explosion in the earth, the plasma temperature is raised to 10^7 from the earlier 10^6 K, necessary for the formation of carbon which is not ferried from the sun to the earth, this carbon with nitrogen of the earth and protons of the solar wind are the ingredients of life on earth and the suspected complete absence of nitrogen on Mars or other solar planets inhibits life on these planets.

I must render my deep gratitude to DTM and its directors since 1948 that it attracted me to do research in earth physics, a life-long interest that has given me great satisfaction in having come, I hope, to their expectations and to solve many issues in earth physics to a working stage.

1
Physics of the Solid Earth

1. Earth and the Solar System

The earth is one of the planets of the solar system. The sun is an average star in the spiral galaxy which also includes the Milky Way. The nucleus of this galaxy is a disc of about a hundred thousand light years in diameter. There are two spiral arms extending from the nucleus. The stars have evolved out of an immense cloud of gas that arose at the beginning of creation or at the time of the big bang. The density blobs and the gravitational pull consolidated the matter into stars, planets and other heavenly bodies. From the observation of the stellar spectra, we can learn about the temperature or the state of turbulence in the star. The temperature rises on account of compression till nuclear reactions begin converting light elements into elements of higher atomic weight. In time, a star assumes the form of a giant red star and later to a red dwarf or a neutron star. The pulsars that emit radio waves are rotating stars. The sun is largely composed of hydrogen which is changing to helium through a series of nuclear reactions at high temperature. The glowing outside atmosphere of the sun is 5700°C. The corona or the luminous envelope around the sun can be seen at the time of solar eclipse. The protons that stream out of the sun constitute the solar wind. Solar flares are observed on certain areas of the sun's surface that emit large amounts of ultra-violet radiation, X rays and fast ions. The flares are noticed specially when the sun also shows dark spots known as sun spots. Strong magnetic fields are associated with sun spots. There is an eleven year sun spot cycle which is also the period of disturbances in the earth's atmosphere.

Sun is the center of the solar system and the planets of the solar system including the earth revolve around the sun. Earth has one satellite, the moon. Mars has two tiny satellites. A group of asteroids also revolves around the sun between the orbits of the planets Mars and Jupiter. A planet moves in an elliptical orbit. The movement is described in Kepler's Laws on account of the gravitational attraction between the sun and the planet. If T is time of planet's revolution and R the average distance between the sun and the planet, the law states that T^2/R^3 = a constant. The constant is unity when T is in years and R in AU (the Astronomical unit AU of distance is the average radius of the earth's orbit). It is 1.496×10^{11} m. The earth revolves around the sun in one year, and it rotates around an axis inclined at an angle 23.4° with the perpendicular to its orbit. The moon revolves around the earth in 27.3 days and also rotates around its own axis at the same rate, thus presenting the same face to the earth. The plane of the orbit of the moon is slightly tilted from the plane of the orbit of the earth or the ecliptic. Meteorites are remains of meteoroids, revolving around the sun, that get caught in the earth's gravity field. Meteorites are mostly chondrites and stones.

2. Distance of Celestial Objects

The distance from the earth can be obtained from observations taken from two positions of the earth in its orbit. The average distance of the sun is calculated by assuming the earth's orbit to be a circle, traversed by the earth at an average speed of 30.34 km per second, in one year. This speed is determined by the aberration observed for a star. When a star situated at right angles to the ecliptic is observed, the speed of the earth is given by the relation: $v = c \tan \alpha$, where c is the velocity of light and α is half the angle between two sites in the opposite passes of the earth in its orbit. The aberration is the annual angular motion about the average position of the star on account of the apparent addition of the orbital velocity to that of light so far as the direction of approach from the front or the back is concerned.

3. Size and Shape of the Earth

The earth is practically a sphere. The poles are the points on the surface where the axis of rotation touches it. A meridian is the great circle passing through the poles. The meridian passing through Greenwich in England is by convention a meridian of zero longitude. The angular distance of the meridian through any place is the longitude of that place. A great circle midway between the poles is known as the equator. The angular distance of a circle parallel to the equatorial plane through any place is known as the latitude of the place.

The surface of the earth shows, volcanism and orogenesis. There happens to be an equatorial bulge on the earth. There are surface undulations and evidence of spalling and scabbing. The equatorial diameter is 43 km greater than the polar diameter. A plumb line is tilted from the axial or the vertical direction. The angle of tilt is equal to the centripetal force divided by the force of gravity. It is in radians, w 2.R sin L/g, where w is the angular speed of the earth and L is the latitude. Knowing this angle of tilt, R can be calculated.

(a) The Earth

Item	Measure
Age	5.6 billion years
Angular velocity	7.292×10^{-5} rad/sec
Mean distance of sun	1.5×10^{11} km.
Semi major axis f earth's orbit	1 AU
Average orbital velocity of the earth	2977 km S^{-1}
Length of the year	3.1558×10^{7} S
Equatorial Radius	6378.14 km
Polar radius	6356.75 km
Radius of the earth	6372 km
Outer radius of the liquid core	3480 km
Outer radius of solid inner core	1216 km
Mass	6.03×10^{27} kg
Mean Density	5.52×10^{3} kg/m^3
Mean Density of Crust	2.8×10^{3} kg/m^3
Area of continents and continental shelves	2.01×10^{14} m^2
Ocean surface area	3.09×10^{14} m^2
Mean height of land above mean sea level	0.84 km
Atmospheric pressure at surface of the sea on the equator	1013.25 mb

(Contd.)

(b) Moon

Mass	7.35×10^{22} kg
Radius	1860 km.
Average Density	2.73×10^{3} kg m^{-3}
Distance of moon from earth	3.844×10^{5} km.
Time of Revolution of the Moon around the earth	27.3 days

Source: The table reproduced from Geophysics (Nanda[1]) gives some earth and moon data.

4. Gravitational Differentiation

During the assembly of the earth, out of material arriving at site in liquid driblets from the sun, gravitational differentiation starts as soon as the material is taken up by the corona that perform the delivery as well as heating it to plasma temperatures. It was hinted by many scientists that the earth should consist of various shell one inside the other, but the concept was never pursued to any conclusion. The outer boundary of the liquid outer core is at 2900 km depth from surface. The author read papers at many UGGI general assembly's regarding the core structure or magnetism of the earth, but a connected account is being given in this book for the first time since only recently he could get convincing evidence when the work of many institutions could only be explained when reliance is placed on his theories. There were hardly any geophysicists grounded in nuclear physics or who found the deficiency of knowledge, of properties of materials at high temperature and high pressures in the earth, as insurmountable for them to tackle. Even the obvious evidence of explosions inside the earth like the presence of tritium in the atmosphere of earth was ignored. The author has found that atomic explosions could occur in a planet. He found evidence of two such explosions in the earth which is clear and strong.

5. Special Status of Earth among Solar System Planets

The proto-sun was already in the main sequence of the evolution of stars when it shed earth and other planets. The sun had gone through the T-Tauri phase which is a turbulent one. When earth was formed, the sun had not yet emerged from the T-Tauri phase. The distinction of the earth will be evident from the relative values of the rotation moment per unit mass (K_m). The earth has the highest value. The group of planets, that have high values, must have formed using material mostly shed in the T-Tauri phase when the rotation is enhanced from magnetic interaction within the star. We also believe that the solar planets included, as considered by ancient astronomers, a planet of small size almost approaching the size of the moon but which was neither seen nor remembered later. In its place we have considered the present group of asteroids that revolve around the sun with orbits within the orbits of the planets Mars and Jupiter. We have denoted it as a shattered planet (SP). It should have a large value of K_m. The planets Jupiter, Saturn, Neptune and Pluto are hybrid, composed of fragments from both phases of the sun. Apart from the distinctive rotation value, earth also has Ruthenium a rare material, which has not been detected in the analysis of other meteorites.

6. Earth's Magnetism

From the magnetism evident for the earth, earth is like a large magnet with axis of rotation of the earth practically along the axis of the earth magnet. The first atomic explosion in the earth would have tilted the earth and paleomagnetism shows earth had its magnetism inclined to the axis of rotation by 11.2 degrees. Now this axis is inclined at 23.4 degrees. This would mean that there was a second explosion to further tilt the axis. This adds reality to the second explosion seen in the enhancement of earth's magnetic field, hinted by the biological observations (Hale[2]), that had occurred about 2 billion years before the present.

7. Internal Structure of the Earth

Information about the internal structure is gleaned from the travel time data of earthquake waves which travel from the focus of the earthquakes to the observation point going along various sections of the structure and undergoing reflections and refractions or conversions and scattering. The velocity of the waves is known from explosion seismology and the density at any place can be derived from the velocity at this point. The suppression of shear waves beyond a certain angular range indicates the presence of liquid deep inside the earth barred for travel of shear waves.

The presence of low velocity layers can also be inferred from the focusing and defocusing effects. The Mohorovicic discontinuity (the Moho) is at the base of the shallow crust floating on a relatively dense but weaker asthenosphere, below the upper 70 to 120 km, known as the lithosphere. The part below the Moho is also known as the mantle extending to the outer radius 3480 km of the liquid core down to the outer radius 1216 km of the solid core. The mantle is composed mainly of silicates and the liquid and solid cores are mainly iron with some mixing of sulphur or nickel. Table below gives various zones inside the earth. V_l or V_P, and V_s are respectively the velocity of the longitudinal and the shear waves. The shear waves do not exist in a liquid medium.

8. Important Past Events Regarding the Earth

(a) The mass of the earth has been delivered from the sun.

(b) Initially the earth was a molten sphere in space with a thin solid crust. Chemical transformations and phase changes produced a solid crust and nuclei of continents. An inner solid core started building up from the center outwards. Few hundred million years after the formation of the earth and the gravitational differentiation of elements and of isotopes of uranium, a thin layer of ^{235}U in the innermost core went critical to have an atomic explosion in the earth. The explosion projected a small piece from the outer layer of solid core

Table : Structure of the Earth

Pressure (Kilobars at top)	*Depth (km)*	*V_p (km/sec)*	*Designation*	*Density (g/cm^3)*
0	0 to 40	Continents 5.5 -7.4	Crust A	2.4 to 3.0
		Oceans 5 to 12	Crust A	
10	40	7.9 to 9.0	Upper Mantle B	3 to 3.5
140	10-1000	9 to 11.4	Upper Mantle C	4.5
350	1000-2700	11.4 to 13.6	Lower Mantle D	5.6
	2700-2900	13.6	Lower Mantle D*	10.0
Density step				
1360	2900-4990	8.1-10.4	Outer Core E	11 to 12
3200	4990-5120	10.4–9.5	Outer Core F	12 to 12.5
3210	5120-6217	3.5	Inner Core G	13-16
	*6217-6339		Liquid Copper	8.94 when unstressed
	*6339-6372		Innermost Core	10.7 when unstressed
	6372		Center of Earth	

Note: The entries marked by star are an addition to existing earth physics from researches of the author. The inner-most core contains elements heavier than liquid copper.

The densities can be calculated if the velocity of P waves there is known. The velocity of P waves is given by the equation $V_p = [(K + 4\mu/3)/\rho]^{1/2}$ and $V_s = (\mu/\rho)^{1/2}$ where μ is the rigidity and K the bulk modulus.

The Moho is in patches, under continents the depth is 40 km and under the oceans, the Moho has depth 5 to 12 km. (The Moho could be the boundary between the lower crust granulites and the ultra-basic upper mantle olivine or pyroxene). The crustal thickness of 8 to 10 km under the oceans and 12 to 30 kms on land is reported on the high side since slight continuous increase of seismic velocity with depth was ignored as pointed out by Nanda during IVGG discussions.

encircling the innermost core. The power of the first explosion will be extremely high as can be judged from the fact that a small planet about ten times the size of the moon which was supposed to have existed between the planets Earth and Mars, became a shattered planet (SP). The swarm of pieces of this SP moves as a group in the sky between the orbits of Mars and Jupiter and keep sending occasionally a meteorite to earth. A scab from the earth's solid core was blown beyond earth's Roche limit (the distance within which any piece gets broken up by the tidal forces and returns by gravity to the large earth in the neighbourhood). The author has estimated from the known speed of the scab away from the earth due to tidal forces as determined by Singer[3] and the present distance of the moon from the earth, that the initial throw took the scab out to a distance like 30 R where R is the radius of the earth. The scab thus provided the substratum close to earth where the moon could be assembled from lighter material from the earth blown up by the nuclear explosions inside the earth. The formation of the moon and state of its magnetism will be taken up in a separate chapter.

(c) This first explosion could also produce giant rifts in the earth so much talked about by the modelers of the tectonosphere model.

(d) Ice ages have occurred when an ice sheet spreads over most of the earth's surface. This has something to do with the variation of intensity of radiation of sun's energy as well as the changes in land and sea configuration on the earth, changes in glacier formation and long period changes in climate. Current stage is the fourth glacial age with cultural level rising from the stone-age. The first ice age known as the Nebraskan whose traces are discernible, occurred about 400 million years ago.

(e) The age of the oceans is much less than that of the earth. It is believed that water was present in almost large surface lakes during the first few hundred million years till most of it got changed to polymeric sheets of dimers and sank to form a sphere of water of about 100 km radius around the shell of liquid copper. It, however, was released by the earth's second atomic explosion that took place around 2 billion years before present. The atomic explosions must have led to considerable orogenesis though geologists usually have many causes of orogenesis and epeirogeny which is operating throughout the earth's past.

(f) Nitrogen in the earth's atmosphere is also a special feature of earth. Nitrogen seems to be so essential for development of life and the absence of free nitrogen on Mars and other planets makes the chance of finding life there rather dim. Mars, temperature is more like that of the moon than that of the earth. It may still be as barren as the moon.

(g) The thickness of the sediments under the present oceans only corresponds to ten million years of the sedimentation at present rate. The concepts of see-bottom spreading and plate tectonics explain this.

(h) Most recent period (573 million years) is

subdivided into 3 eras. Table below gives various divisions.

Table: Eras, Periods and Epochs

The Beginning of the Epoch Time in Million Years B.P.	*Remarks*	
Cenozoic Quarternary Pleistocene	2.5 – 3.0	human beings
Tertiary Pliocene	10	
Miocene	25	
Oligocene	40	
Eocene	55	
Paleocene	70	Mammals
MESOZOIC Cretaceous Late	100	
Early	136	
Jurassic Late	155	
Middle	175	
Early	190	redwood
Triassic	225	oysters, birds
Permian	280	dinosaurs
Pensylvanian	325	
REPTILES AND INSECTS		
Paleozoic Carboniferous		
Missisipian	345	
Devonian Late	360	
Middle	370	
Early	395	
Amphibian Silurian	430	
Ordovician Late	450	
Early	500	
Vertebrates		
Cambrian	570	
Cepheloids		
PROTERZOIC No fossil late	1600	
RECORDS Middle	1900	
Early	2600	

9. Plate Tectonics

The earth's lithosphere is made of many plates. Fifteen plates, approximately 60 km thick can be delineated.

The plates are made of the material that arises out of the ridges and the plates disappear under the trenches or collide with neighbouring plates. The continents are riding on plates that collide and give rise to earthquake activity along the Benioff zones or the continental margins. The depth of focus of the shallow earthquakes is less than 60 km. The deep earthquakes, however, are at 700 km depth. In addition there are intraplate earthquakes on account of volcanic activity. Certain hot spots exist (Wilson[4]) in the deep mantle and these cause intraplate eruptions giving rise to sea mounts and guyots. The line of such guyots defines the line of motion of the plate above the hot spot. The motion of a plate relative to the mantle can be determined from the traces of volcanoes. Dietz[5] has explained the ideas of sea floor spreading in support of plate tectonics. Only along the margins of the plates, active thrust faults are visible, e.g. the collision of the Indian plate with the Eurasian one along the Himalayas. In Tibet, the normal faults are approximately north-south and the tension axis are east-west. The position can be modeled by a "dead triangle" in a plastic material indented by a rigid plate from the south and tensile pulls at right angles. Around this triangle, the Indian and the Chinese plates are pushing against each other.

The continents ride on crustal plates that move and rotate and push against each other. There is subduction of plates and a portion of a plate may even override another subducting plate. The plates move relative to each other at a speed of few cms per year. The plate boundaries are sometimes the same as the continental margins or ocean ridges constitute zones of orogeny and earthquakes. Davis and Solomon[6] discuss the variation of plate speeds from the Cretaceous era. The cause of the plate motion must be the large sized convection cells below the lithosphere which drag the plates along. The absolute motion of the Indian plate since the late Cretaceous was anomalously fast, being of the order of 15-20 cms per year. The details of plate motions are still controversial. Belousov[7] for example favours that the entire continent may not move and the motion could be restricted to relatively smaller cells as in orogenic events. Bhattacharya *et al.*[8] describe evidence of sea floor spreading in India.

The continental crust consists mainly of sial having abundance of silicon and aluminum lying over a deeper layer of denser material sima which has greater proportion of magnesium in place of aluminium. Under the deep sea, sima is close to surface being covered only by a few kms of light sediments or basalts.

10. Rifts and Trenches

The rift system that runs over the Atlantic ridge and continues in a branch to the Siberian coast is also the locus of many earthquake epicenters. Ridges and rifts are also associated with gravity anomalies. The unusual flow from the rifts indicates an ascent of magma masses from the deep oceanic crust. The rifts by themselves cannot accommodate the expansion of the earth's surface and there must occur transform faults as discussed by Allen[9]. Recent seismic activity is often confined to short part of the fracture zones. The faults connect two expanding ridges. A transcurrent fault will offset a ridge. The oceanic rifts being rupture rifts, is supported by paleomagnetic observations. Gravity anomalies are associated with transform faults.

Similar in many respects are the deep sea trenches. These too are deep narrow rifts going down to about 5 kms from the sea bottom. The trenches are only 5 to 10 km in width. They also show anomalies with regard to isostasy and heat flow. Sometimes, there is a continuation of the trench over land. The trenches are short, while the oceanic rifts are worldwide. The trenches run parallel to the island chains and lie on the convex side of the island arcs. The depth of an ocean trench may exceed 8 km (Mariana trench). The oceanic trenches and the island arcs are characteristically concave towards the overriding plate.

11. Epeirogeny

There is constant uplift or sinking going on over large areas of the crust. This happens because of plastic instability below and presence of convection cells there. The rivers change their course partly because of epeirogeny. The configuration of the oceans changes when the continents drift, break up or unite. The Paleozoic breakup of the giant Pangea continent gave rise to the Atlantic ocean, the Arctic ocean and the Indian ocean. These oceans have expanded in later epochs. Epeirogeny is the main reason that permanent training of rivers for fixed course is not possible.

12. Trace Elements and Magma Genesis

We are digressing into Geochemistry and here only some important conclusions will be highlighted. The interaction between magmatic especially sialic rocks are important in altering primary composition. From data on the diffusion of elements in basalt and obsidian liquids and from the sketchy data for diffusion in silicates, some results are available. The diffusion coefficients for elements like Ca, Sr, Ba, Co, Eu and Gd in a basaltic melt would be in the range of 10^{-6} to 10^{-8} units/sec. The diffusion coefficients of Sr and Ba in obsidian melts are lower by a factor of 5 to 10. Elements such as Sr, U, Al and Ni in the mantle minerals have coefficients in the range of 10^{-11} to 10^{-13} units/sec. The effect of pressure increase is relatively small. The kinetic constraints at temperatures between 1150 to 1250°C would be as given below:

(a) The exchange of a trace element between two 1 cm grains in contact will take about 0.5 m.y. The separation of the grains by 1 cm of similar material will increase the equilibrium time to 1 m.y.

(b) The equilibrium between solids in the presence of melt: if the melt wets the grain boundaries, equilibrium can occur in less than 0.04 m.y. If the melt occurs in interconnected tubes, times may increase to 0.5 m.y.

(c) Equilibrium within the melts: The characteristic transport distance will be 5 m in 1 m.y. and 5 km in 1 b.y. Thus we may expect all phases in partially molten mantle to be in elemental equilibrium. It is unlikely that significant melt separation can take place. A large (2 km) magma chamber will reach solidus temperatures in about 2000 years. The conclusion based on kinetic considerations is that in magmatic circumstances, an equilibrium approximation is satisfactory.

13. Isotopic Tracers

Some light elements like H_2 or O_2 have minor isotopes D or ^{18}O. The ratios between these minor isotopes (D/H_2, $^{18}O/^{16}O$) vary in nature.

Taylor and Epstein pointed out in 1962 that these isotopic variations can be used as petrogenic tracers. Pb isotopes are particularly useful. Isotopic tracers are most efficient as tracers when there is large contrast between different environments. Rb/Sr ratios are markedly higher in the continental crust than in the mantle, so the interaction between these two reservoirs are best traced by Sr isotopes. It is important to use a combination of as many isotopic systems as possible for understanding complex petrogenic processes. It is a separate science in itself.

2
Petrogenesis of Rocks

1. Source Determination

Trace elements are helpful in determining some characteristics of the source. Basalt derived from mantle sources are heterogeneous in composition. Alkali basalts show relatively constant heavy rare earth abundances. Mid-ocean basalts have variable rare earth abundances. Ridge basalts would be derived at shallow depths less than 30 km within the plagioclase as a residual phase following melting. Petrogenesis of granitic rocks is more difficult to understand. Granites could be the product of long fractionation of mantle derived melts or from remelting of sediments or gneisses. The Rb/Sr ratio of crustal rocks (granites and sediments) is far higher than the ratio for the mantle material, (3 compared to 0.03). The ratio $^{87}Sr/^{86}Sr$ grows at a rate 10^{-4} per million year compared to the maximum growth rate of 10^{-6} per m.y. in a mantle reservoir. Granite formed by the re-melting of the continental crust will, therefore, have a higher strontium isotopic ratio than the one derived from the mantle fractionation (Hurley *et al.*[10]).

The basaltic liquid regarded as primitive could not be in equilibrium with mantle material at the depth where partial melting occurs. The plate genesis and subduction hint at a rock cycle where the igneous, metamorphic and the sedimentary rocks get converted from one to another and mix.

2. Hydrothermal Circulation and Petrogenesis

Taylor[10] pointed out in 1978, the significance of water-rock exchange processes and the existence of low ^{18}O magma types. Juvenile CO_2 – H_2 rich fluids control the conditions of partial melting at various depths. Partial melts generated in the mantle rise as a diapir or as rapidly moving segregated melt. In volcanism or continents above the subduction zone, magma is produced within the mantle. The collision of two continents results in thrusting and deformation. The present convergence in the Himalayas 1.5 cms/year is less than one half of the convergence between India and Eurasia which gives rise to folding. The continental crust grows in time at the cost of the mantle. The oceanic sediments are returned to the continents at the subduction zone. At the margins of the continents, basins are formed and the continental crust shows slow extension outwards through the uplifted ridges. Most basalts have undergone some fractional crystallization and primary magmas are rare. Fractional crystallization process is less important for basalt and granite petrogenesis than partial melting.

3. The Deccan

The Deccan basalts cover almost five-sixth of the peninsula. The thickness ranging from 2000m in the north to 500 m in the south and 150 m in the east. The radioactive ages of samples of basalt are approximately 60 m.y. at various sites. The eruptions responsible for the spread of these basalts must have been from fissures since there are no relics of volcanic cones. From age determination. The formation of the basalts started about the time the Carlsberg ridge was formed in the Indian Ocean. The west Godavari depression is characterized by gravity low (–40 mgal.) Ridges, terraces show up as highs. In general the crust has a two layer structure (2.5 to 2.7 g/cm^3 in the upper 10 km zone and 2.9-3.05 g/cm^3 at greater depths.

The inter-plate compressive forces are activating some of the faults. The average geothermal gradient in the peninsula is of the order of 20 to 25°C/km (Rao and Rao[11]). Several hot springs with water at 55 to 75°C suggest that ground water must be penetrating to a depth of 3 to 4 km. The west coast of India has undergone appreciable vertical movements during the tertiary. There has been formation of sedimentary off shore basins. Earth's interior can also be probed by geomagnetism. The anomalies of the field are related to the structure of the crust. Mantle is the seat for secular variations.

3
Geophysical Exploration and Geo-Chronology

1. Electrical Methods

For the exploration of water and electrically conducting minerals beneath the surface of the earth, electromagnetic induction or the electrical conductivity methods are useful.

In the Wenner array, four electrodes are equally spaced along a straight line; the array spacing is 'a'. The electrical current I is fed into the earth between two electrodes and voltages are measured between two other electrodes in the neighbourhood. Resistance $R = \mu\Delta l/A$, where μ is the specific resistance, and Δl is the length and A is the area of cross-section.

We get grad $V = \mu I$, where I is the current density per unit area. At a distance r from a current source sending the current into the earth $I = i \times 2\pi r^2$ since the current is bulging out downwards, the potential gradient is $-dV/dr = \mu I/2\pi r^2$ which enables one to calculate the potential distribution in a homogeneous conducting semi infinite medium.

In the Wenner array, four electrodes are equally

spaced along a straight line, the distance 'a' being the array spacing. In the Schlumberger array, the distance between the mid-points of the current and the potential electrodes can be varied. If this distance x is zero, we have the symmetrical Schlumberger array for which the specific resistance is simplified. The aim of resistivity survey is to delineate the boundaries in heterogeneous ground. In the field, resistivity surveys are carried out by using a high tension battery pack as a source of current, four metal spikes for contact with ground and connecting wire and ammeter. Now-a-days most measurements are made using alternating current ac at 4cps, low frequency a.c. current suffers less from skin effect. For a.c. the depth of penetration is $h = 503.8\,(\mu/n)^{1/2}$, where n is the frequency, Vertical structures such as faults, dikes, contact zones can be shown on the resistance profiles across the strike of these features.

2. Geochronology

Geochronology deals with the determination of ages of various strata from the study of various rocks and fossils, if any. The fossils can help in placing the relative ages of the rocks, but absolute age can be determined from the study of radioactive decay products or relative proportion of isotopes of suitable elements. Radioactive material acts as a natural clock placed in the rock at the time of its formation. If n is the number of radioactive atoms at anyone time, the number will be reduced at the rate $dn/dt = -\lambda t$, where λ is the decay constant. The decay constant is derivable from half life often available in literature. The mineral most commonly used for Uranium based dating is Zircon (Zr Si O_4). Zircon occurs as accessory mineral in many rocks. Uranium atom when replacing few of the zircon atoms remain entrapped through metamorphic events.

Actual comparison of the isotopes of lead minerals such as galena of different ages revealed that ^{206}Pb, ^{207}Pb and ^{208}Pb have indeed increased in quantity with time. The diagrams known as the Concordance Diagrams help in proper age determination even when some of the material has been depleted. Lead isotope analysis has led to the determination of age of the earth as approximately 5 billion years.

3. Rubidium-Strontium Method

^{87}Rb atom decays to ^{87}Sr emitting a β-particle. The common isotopes of strontium are ^{87}Sr and ^{86}Sr. ^{87}Sr initially present is known from its natural isotope ratio and the present quantity of ^{86}Sr. The added ^{87}Sr is from the disintegration of ^{87}Rb and this isotope of Rb is also measured in addition to the total amount of ^{87}Sr. The time elapsed t can now be calculated from the equation:

$$(^{87}Sr/^{86}Sr)_{total} = (^{87}Sr/^{86}Sr)_{initial} \text{ plus} (^{87}Rb/^{86}Sr)\,(e^{\lambda t} - 1)$$

The Rb–Sr technique is useful in pursuing the history of the rocks. Suppose that there are three rocks A, B, and C, differentiated from a common magma T years ago with a common initial magma ratio of the two isotopes, but after a possible metamorphism t years ago, the three initial ratios will become different, but the total ^{87}Sr of the rock is expected to stay the same since it does not migrate out. In the course of time T, the ^{87}Rb was depleted by decay and ^{87}Sr was enriched correspondingly so that the isotopic composition of the rocks migrated from A, B, C to A', B', C' respectively. The line through A', B', C' is the T isochron indicating the variation in the total ratio of the strontium isotopes for the whole rock and its slope is $(e^{\lambda t}-1)$ or almost λT from which T is determined. T being the time in the past when the rocks were rehomogenised. The intercept gives the initial isotope composition of strontium, T years ago if the rocks were all reheated during a metamorphic event at time T (let) in the past, the minerals were changed somewhat giving a new starting point for Sr evolution in each of the rocks. But T isochrone would not be affected. And T the age of the original isotopic ratio of the parent magma can be determined.

The Rb–Sr method can be used for common rock forming minerals as micas, and all types of feldspars including orthoclase. The great advantage is that the method represents a solid-sold system and there would be no loss of parent or daughter from the system. It has been found very valuable method for dating shield areas of pre-Cambrian rocks. The radio carbon method on the other hand is based on radioactivity of isotopes which have much shorter life time.

4
New Perceptions in Earth Physics

1. Introduction

Earth science needs a fresh look at the problems it faces. There are claims that an earth dynamo based on convection currents inside the earth may explain both the magnetic field of the earth and its reversals. Will then there be a different theory for the emergence of a field on the surface of Mercury which has no possibility of convection currents in it. Similarly, there is emergence of helium from inside the earth. This helium escapes into space after a long residence period in the earth's atmosphere. For reversals and repeated establishment of field considerable energy input must be there over what comes from sunshine. The estimate of James[11] is that an input of energy of more than 10^{11}W is necessary. The growth of the moon so near a gravitating body like the earth was a mystery.

Keeping these blocks in the progress of earth physics in mind, Dr. M.A Tuve selected me after my post graduate Carnegie Fellowship in his Department of Terrestrial Magnetism (DTM) at Washington D.C. and gave me this task of working on earth physics problems, and assured my continuing such work when I return to India after the completion of training in USA. The DTM sent along with me some equipment for my research in microseisms and oceanography that I had started in DTM. My preliminary studies had already impressed him. My theoretical ideas alone could explain some observations of other scientists abroad when the experimenters had no logical explanation. This book deals with modern geology as well as new factors which have been introduced in Earth Physics.

The new factors mainly involve three aspects:

(a) ^{235}U, a spontaneously fissioning element is present from start in all planets and invariably causes atomic explosions in planets,
(b) planetary magnetic fields are not convection driven.
(c) The moon has grown out of planetary explosion driven out scabs and stones from the earth.
(d) I have also discovered that earth has a special position among solar planets.
(e) Physics of solar corona that delivers planetary material in driblets to planet sites is very relevant.
(f) I have also something to say about the earth being the only planet ever likely to grow human civilization
(g) Our existence on earth is likely to be disturbed by the travails against ice ages more than by

problems from atmospheric or ocean storms or earthquakes than by any violent impacts with other heavenly bodies for the remaining life of earth or of the sun.

The planetary physicists need not feel guilty when they find that most planets in their early history had atomic explosions. The presence of a large moon close to earth is only possible if a substratum for its growth was exploded out of the earth to exist beyond earth's Roche limit. This new approach also predicted an iron intrusion in the moon. Study of surface magnetism has been attempted by an electron reflection method by Anderson and McGuire[15] and they have yet to decipher the experimental results. Hundreds of miles long track of a magnetic anomaly is visible starting from Rima Sirsalis feature on the surface of the moon as if an extremely long magnetic rod has been thrust into the moon from outside. In our view, the planetary explosion had projected a long iron scab beyond the Roche limit from the earth and the anomaly only reflects the trace of this rod inside the moon. Another gross deficiency in present earth physics is continued dependence on a geodynamo based on convection currents inside the earth for explaining terrestrial magnetism. For most of the geophysicists, the geodynamo hypothesis was still born because of the stable layered structure discovered in the outer liquid core. The viscosity conditions also were not satisfied as pointed out by Lumb and Aldridge[16] in 1991. We next explain the genesis and the importance of the Earth's Innermost core.

2. Earth's Innermost Core

This is also the title of my first important and now a source paper in earth physics. Here we give a bit improved text of this paper which also has been published in my booklet of the same name in 2004 by the Concept Publishing Comany Pvt. Ltd., New Delhi. The earth initially was a liquid sphere in space with a thin solid crust. There was gravitational separation as various elements settled to positions determined by relative density at the core conditions of temperature and pressure. Liquid copper is the heaviest element (not iron) and so it will go closest to the center of the liquid earth. We have approached the structure problem as of shell formation of decreasing density as one goes outwards from the center. A report by Ramamurthy in the encyclopedia of Geochemistry and Environmental Science has shown that all elements will be in atomic state at core conditions. There are quite a few elements that have melting points higher than the core temperature reached by iron. These will remain as a solid mixture and constitute an innermost core of the earth. For the size of the liquid ball of the earth and the parts per million of the earth forming chondrites and irons, the mass of any element as ppm of the liquid ball can be determined. The total mass of the innermost core was thus found to be 0.943 ppm of the liquid mass. The innermost core must be floating in liquid copper which surrounds it and so it can have a natural vibration, the frequency of which is determined by the theoretical relation between the period of the vibration and the radius *r* of the spherical vibrating body given by Ari Benhem and Sarvajit Singh[17]. In an innovative theory, a guessing step is normal and is the root of all scientific new theories. My contribution is the right guess. I had gained enough confidence by working with Dr. M.A. Tuve the Director of DTM in Washington D.C. USA where I learnt directly from his staff my geophysics and still regard DTM as my alma mater for oceanography and geophysics. I recalled from my membership of the IASPEI Commission on Microseisms that the microseisms recorded by Holcomb[18] of anomalous long period (26 seconds) had received no further notice of geologists or oceanographers. I thought of these vibrations to be the natural vibration of the innermost core floating in fluid copper at core conditions. Starting from the innermost core, further shell structure could be worked outwards from the center of the earth. The spherical radius r of the innermost core and its natural vibration when floating in copper is calculated from the theoretical relation from the book by Ari Benhem and Sarvjit Singh (*loc cit.*[19]):

$V = (2\pi/T)(r/2.501)$. For T = 26 S, the equation reduces to $r = 10.35\,v$, where v is the velocity of shear waves in the liquid. The velocity v of the shear waves is given by the square root of (shear modulus/density). Thus $v^2 = 34.8\,x^2/10.8x$ where 34.8 units is the shear strength of the unstressed material of the floating core of unstressed material of density 10.8 calculated as for a mechanical mixture of the constituents, and x is the compression *in situ*. We have two equations involving r and v. Solving these, we get $r = 18.35\,x^{1/2}$. Knowing that shear wave velocity near the center of the earth is 3.2 km/sec, we get $x = 3.15$ and $r = 33$ km. The innermost core is isolated from the inner solid iron shell by a liquid copper shell and from the quantity of copper brought in by the chondrites and irons, the copper shell extends from the inner radius 33 km to 155 km outer radius. Thus we arrive at the description of two parts of the earth's deep core. In case of earth, an atomic explosion in its early history is confirmed by the detection (Anderson *et al. loc cit.*[15]) in 1977, of a large magnetic anomaly as if a metallic rod, which was incredibly long, had got inserted from outside into the moon.

3. Atomic Explosions in Planets

In the millennia after the formation of the earth, the gravitational sorting out would extend to the separation of a thin layer of ^{235}U on the surface of the uranium layer and when this thin layer has a critical thickness of about 30 cms, a spontaneous explosion would take place. Such an explosion leaves a large number of helium atoms including ^{3}He atoms, from the tertiary fission of some of the fissioning uranium atoms. This is the origin of the helium oozing out of earth at various places including some in east India which was used for supply helium gas and Atomic Energy Commission studied this source. The atomic explosions in planets is a foregone conclusion when we see so many prograde satellites presumably the result of such explosions. In the case of the earth, an atomic explosion early in earth is practically confirmed by Anderson as stated earlier. According to the Greek astronomers, there was a planet about ten times the size of the moon between the orbits of Earth and Mars but it is not there anymore; the Bode law also held that such a planet should have existed. Perhaps it did exist but was destroyed by its own first atomic explosion. We shall refer to this shattered planet with symbol SP in our account. Could the swarm of meteorites that revolves around the sun between the orbits of the Earth and Mars be the remanents of the SP. Mars itself had two quick explosions exploding two tiny moons from within itself, since it started its build up with the central sphere of uranium already in the liquid state.

The magnetic field of the earth became two fold suddenly about 2 billion years before present. This is a possible consequence of a second explosion in the earth. An atomic explosion could enhance the magnetic field. Now that the geophysicists have buried the concept of a geodynamo, we must look for the real cause of planetary magnetism. Since magnetism shows reversals, it has to be from paramagnetism of the content of solid iron in the earth and the magnetism could be caused by a reversible exciting field. A natural exciting field is provided by the relic of the space magnetism, about 3 nano-Tesla in strength. Such low intensity field could not produce the fields that exist near planets since the permeability for low fields is very small. The reason for large value of H for the earth will be given in due course when discussing magnetic fields near planets. The doubling of the field after the second explosion (Hale *loc. cit.*[2]) gives the hint that magnetism must have been on account of certain particles having a magnetic spin possibly due to nuclear magnetism like atoms of tritium which are held captive in a cavity inside the massive solid iron shell.

4. A Discussion about the Ages

The estimated age (determined by Wilde *et al.*[19]) of the oldest rocks on the earth, the detritus zircon is 4.4 b.y. The age of the lunar oldest core is also 4.4 b.y. as determined in 1988 by (Carla and Langmair[20]). This points out to the simultaneous start of the solidification of the earth as well as the moon following the explosion in the earth that started the assembly of the moon beyond the Roche limit (2.5 times the Radius) of the earth around the substratum provided by the scab from the earth.

The age of the meteorites in space is known from the study of the spectral lines. Following the U-Pb method, the age, from the study of rock samples originating from the largest meteorite is 4.56 b.y. The swarm of theses meteorites must have started at the time of the formation of the earth. The planets started their individual growth at about the same time known as the age of the earth. The earth would have taken about 50,000 years for getting to full size as estimated from numerical-experiments. Lyttleton[21] has used this figure in his analysis of the earth-moon sub-system.

5. Liquid Copper Shell

Liquid copper shell around the innermost core,

assumed to be stable for simplicity of treatment in spite of the density inversion with respect to the larger enclosing iron shell which extends to 1216 km radius while the outer radius of the copper shell is only 155 km. the lower density materials are existing as at least dimers to get higher density material to suit the position of the shell. The presence of liquid copper shell and the solid innermost core clarify the observed higher values of the velocity of the P waves (M. Ishi and Dzievoski[22]) and there need not be any introduction of biaxial crystallization of the solid iron core, particularly because the inner core will be in liquid state. At high temperature iron will be in the liquid state.

6. Temperature of the Innermost Core

(a) Useful Data

Decay Constant f.

The temperature of the outer edge of the solid iron shell is 6000 K, i.e. below the melting point of iron under core conditions. The compression at the inner boundary is density *in situ* divided by unstressed density or 14/8 = 1.75. For the speed of S waves of 3.5. km per sec, the rigidity of the core material is 208 $\times 10^{10}$ d/cm^2 and the bulk modulus is 416 d/cm^2. The value is practically of 166 units for unstressed iron multiplied by square of compression. This is a confirmation of our empirical relation between rigidity and compression. The temperature coefficient of decay (*f*) is given by equation $K_T = K_0 \{1 - f(T-$

$T_0)\}$. For iron f is 2.2×10^{-4}. When there is compression x, the unstressed decay constant requires multiplication by x^2 from the empirical relation $K_{T,P} = K_{0,0}\, x^2$. The justification is from the physics of continuous matter. Thus, the applicable decay constant is known.

(b) The calculation of compression of the innermost core of radius R immersed in a liquid can be calculated from the theory given by Sarvajit Singh and Ari Benmenahem (*loc. cit.*[19]). R is related to the velocity v of shear waves in the body of radius R.

$$v = (2\pi/T)\,(R/2.501)$$

The innovative step taken by the author was identifying the vibration with the microseisms of 26 second period recorded by Holcomb in 1983 (*loc cit.*[18]). Using above equation for v and period T = 26 s of vibration of the composite material of the innermost core, a mechanical mixture of known proportions. Their combined ppm value related to the entire mass of liquid earth was 0.943 and rigidity contributed proportionally is 66.7×10^{10} d/cm^2, and density is compression (x) times the unstressed value of the mechanical mixture or 10.7 × "x", the compression. Thus we get $v = 1.79\, x^{1/2}$. Using this v, the equation for R now becomes $R = 18.35\, x^{1/2}$. Thus we determine R = 33 km and x = 3.15. for the innermost core or for the surrounding liquid copper. The compression can also be known from the change in material from iron to copper from the relation $x = 1.75 \times 416/66\, x^2$ or $x = 2.2$ at the temperature of 6500K of iron close to center. (We have used the bulk modulus of the composite material from the tabulated proportions in ppm values relating to entire earth reaching the center without melting.) The compression experienced by the innermost core at the center has to be more because of the expansion with rise of temperature above 6500 (to the melting point of iron under core conditions at center) as well as the fall in bulk modulus. This is the argument for the changed value of compression. The compression $x = 2.2/[(1.081 \times 10^{-4} x^2 . \delta T)/x^2)]$. Use $x = 3.15$ and $\delta T = T' - 6500$. We get, $\delta T = 349$. Thus the temperature of the innermost core T' is 6849 K.

7. Heat Outflow from the Innermost Core

The innermost core is a repository of radioactive or fissioning atoms. It must be giving out heat. Since there is no source or sink of heat in the copper shell, the temperature of any spherical surface of radius r in it would be $T = a + b/r$. So that heat outflow at all radii is the same. The constant b can be calculated from boundary conditions. $b = 1.501 \times 10^7$. The heat conducted across any spherical surface Q would be $4\pi\, b\, q$ where b is the conductivity of the liquid copper, where the compression is 1.9 (from changed material, copper in place of iron). which makes the conductivity 1.9×354 where 354 units is unstressed conductivity. Thus we get Q, the heat conducted across as 1.25×10^{11} W. See discussion by James[23].

5
Magnetism

1. Magnetic Field

Magnetism is bipolar in nature, each magnet having a north or positive pole and a south or negative pole. The force between the magnetic poles is given by a law analogous to Coulomb's law for electrical charges. The force experienced per unit pole strength is the field intensity. The force is proportional to the pole strength and the field intensity. The flux density B is proportional to the magnetizing field H or $B = \mu_R H$ where μ_R is the relative permeability of the medium. The permeability of vacuum is $\mu_0 = 4\pi \times 10^{-7}$ in SI units. The flux density can be measured in voltseconds/m^2 or in Tesla which is 1 weber per square meter. The force acting between two poles *m* and *m*'ata distance *r* from each other in a medium of relative permeability μ_R is $(\mu_0/4\pi)\ (1/\mu_R)\ (mm'/r^2)$. In cgs units, the unit of magnetic intensity is Gauss which is equal to 10^{-4} Tesla or 10^{-4} weber per meter square. A sub-unit γ is more convenient, it equals 10^{-5} Gauss.

2. Induced Magnetization and Bulk Susceptibility

The magnetization produced in rocks by the locally acting external field is called induced magnetization which depends on the bulk susceptibility K_b of the rock body. When a rock specimen is put in a weak external field H_{ext}, the induced magnetization in the bulk is $M = K_b H_{eff}$ where H_{eff} is the effective field = $H_{ext} - H_D$ where H_D is the demagnetizing field due to induced magnetism. In most of the rocks, the demagnetizing field is insignificant. The bulk susceptibility of rocks depends not only on the content of magnetic minerals but also on their grain size and shape etc. The susceptibility is strongly dependent on the field strength and the past history of magnetization (hysteresis). Note that large grains have higher susceptibility and lower coercive force. The hysteresis curve shows the variation of the induction B with H_{ext}. The field H_c is the coercive field for reducing B to zero. The magnetic flux B_R is the remanent magnetism.

Because of hysteresis, the bulk susceptibility of a rock sample is dependent on the state of magnetization and to some extent on the non-intrinsic parameters of grain size, internal stress etc. The hysteresis effect is roughly represented by the coercive force H_c which is 10 to 1000 oersteds for most of the rocks.

The ratio of the bulk moment to the effective field gives susceptibility. The effective field is $H_c - H_D$ where the demagnetization field is given by J_i, the induced magnetic pole strength. The demagnetization factor is 1/3 for spherical grains. The bulk susceptibility is

$K_b = p. k. (1 - N_R k)$ where p is the volume fraction of the magnetic material and k is the intrinsic isotropic susceptibility. Anisotropy of the magnetization if any is measured by the determination of the magnetic moment in three mutually perpendicular directions. The anisotropy of the susceptibility suggests gravity foliation or lineation in the direction of flow. In sedimentary rocks, the action of gravity produces a foliation parallel to the plane of bedding ($K_{max} - K_{min}$ plane). In the case of the metamorphic rocks the observed fabric is used to estimate stresses since there is no obvious lineation or foliation. In case of igneous rocks one determines the fabric of lavas and the history of flow. Volume susceptibilities of some materials are given below:

Mineral	*Volume Susceptibility, (to convert these values to electromagnetic cgs units, divide by 4π)*
Granite (without magnetic mineral)	10650
Granite (with magnetic mineral)	20,400
Basalt	5,000 – 1,200,000
Magnetic ore	700 to 1.4×10^8
Rocksalt (diamagnetic)	–100
Ice	–90

The magnetic field is measured by a fluxgate magnetometer or a proton magnetometer. The proton magnetometer is based on the Larmour precession experienced by protons in the field F. The Larmour frequency is w = gamma × F, where gamma is the gyromagnetic ratio for the protons. The field strength is calculated from the frequency. The diurnal variation has practical significance in magnetic surveying and mineral prospecting.

The long term changes in the field are known as secular variation. The earth's magnetic field is composed of a main field and a small field which varies rapidly and has its origin outside the solid earth. Small variations in the main field constitute the anomalies of interest to the exploration geophysicists. The position, of the north magnetic pole of the earth as recorded at London seems to wander in a closed loop, possibly closing in about 500 years. A large part of the earth's field can be represented by an axial dipole. Some non-axial dipoles, not at the center of the earth, would also be required to explain the detailed field all over the globe. There are large local anomalies of the non-dipole field. For example an anomaly of about 16,000 gamma over thousands of square km must be due to a deep source at the core boundary. Westward movement of certain anomalies has been explained by the waves in the convection flow of the dynamo theorists or by differential rotation of some shells of the earth structure. The subject is still open for suitable theory.

3. Diamagnetism, Paramagnetism and Ferromagnetism

A substance is called diamagnetic, if the magnetic susceptibility k is negative; for para- and ferromagnetic substances, it is positive, i.e., a magnetization is induced when a magnetic field is applied. In diamagnetic materials, the induced magnetization is in the opposite direction to the applied field, these have negative susceptibility of the order of 10^{-5} SI.

Paramagnetism is dominant in atoms with odd number of valence electrons. The susceptibility of paramagnetic substances decreases inversely with absolute temperature. Paramagnetic minerals found in natural rocks are pyroxenes, olivines, garnets, biotitic and amphiboles. In ferromagnetic materials, the susceptibility is several orders higher, sometimes more than 10 SI. It is dependent on the temperature and on the strength of the applied field. The coupling of spin moments of electrons leads to spontaneous magnetization. The alignment of electron spins occurs within 10^{-6} m. The examples are Fe, Co, and Ni. In ferromagnetic substances, the spin moments are arranged in two unequal sub-lattices. The spontaneous magnetization of ferro- and ferri-magnetic substances decreases with increasing temperatures and disappears at Curie point ($T_c = 580°C$ for magnetite). In anti-ferromagnetic materials, there are two equal anti-parallel sub-lattices resulting in zero magnetic moment. An increase in temperature would make such materials more magnetic, but the magnetism drops to paramagnetic values above Neel temperature. An anti-ferromagnetic mineral is hematite. Practically all constituents that that give high magnetization to rocks are ferromagnetic minerals. These minerals can be divided into two group's iron-titanium-oxygen group and the iron-sulphur group. The geochemistry of the first group can best be described in the ternary system

$$FeO - Fe_2O_3 - TiO_2.$$

4. Non-Axial Field and Curie Point under Core Conditions

The field for which the generating dipole is non-axial had intrigued the dynamo theory. The portion of the magnetic field that does not reverse can easily be ascribed to some projections on the solid iron core whose tips are at a temperature below the Curie point on account of their dipping in cooler fluid that gives them ferromagnetic susceptibility and the reversing exciting field may not be above the required coercing field on account of the position of the iron tip on the hysterisis curve. The entire solid iron core must be just above the Curie point except for certain tips projecting into the cooler liquid.

Theoretically, the fcc γ iron is suspected to have a Curie Temperature of 1800 K which will rise with pressure. If one could make an extravagant extrapolation, this stable form might be re-occurring giving the Curie point between 6400 and 6000 for core conditions so that iron stays above the Curie point.

5. Displacement of the Dipole from the Center of the Earth

It could be due to the non-central location of the helium reservoir after the massive atomic explosion in the innermost core. The centrally displaced exciting field will create paramagnetic moment in the core which will have cylindrical symmetry around the dipole in the helium reservoir and not around the axis of rotation.

6. Remanent Magnetism in Rocks

Most common rocks of the upper part of the crust have a remanent magnetism J_r in addition to the J_i induced by the earth's field. The intensity of J_r is particularly large for in the igneous and, thermally metamorphosed rocks, for which the Koenigs-berger ratio $Q = J_r/(K_b.F)$ is often greater than unity and sometimes exceeds 100 and gives rise to magnetic anomalies. J_r ranges from 10^{-4} A/m for weekly magnetized sediments to 10 A/m for basic igneous rocks. The Q values for some of the materials are as follows: Basalt (1 to 160), silts (2-3.5), Gabbro (9.5) Granite (0.2-0.9), sediments (0.1). The magnetization (vector sum of J_r and J_i is found to be nearly parallel to the present geomagnetic field). The resultant magnetization can be expressed as $J_e = K_b H_{eff} \pm j_r = K_e H_{eff}$. The natural processes by which remanent magnetism is acquired by rocks will be listed next: DRM or the depositional remanent magnetization, this is acquired when grains settle in still water and get embedded in sediments. DRM is relatively stable. TRM or thermo-remanent magnetism acquired during cooling from Curie point to room temperature in the presence of a magnetic field. It will apply to baked pottery and has become a tool of archaeology. Isothermal or IRM is important only for rock outcrops that might have been exposed to intense local fields for example by a lightning stroke. Piezo-remanent magnetism arises due to application or release of pressure, e.g. as an indication of an impending earthquake.

7. Paleo-Magnetic Field

Earth's paleo-magnetic field, i.e. the field, in olden times when the remanent, magnetism was created in a rock sample, can be determined from the age of the rocks besides the magnetic moment of the sample. Paleo-magnetic field gives an idea of the migration of the magnetic poles and drift of continents and it is a powerful tool for tectonic studies. Oceanic rift systems have been studied in various submarine surveys. The rift system that runs over the Atlantic ridge and continues in a Siberian branch is also the locus of many earthquake epicenters. Ridges and rifts are associated with gravity anomalies and a deviation from isostasy. The unusual flow of heat from the rifts indicates ascent of magma masses deep below the oceanic crust. The Atlantic rise must have been formed by the ascent of magma along the rift that runs along the ridge axis. There are basalts present along the rise and the age of these basalts is only about ten million years. The rifts by themselves cannot accommodate the expansion of earth's surface without relative movement along transform faults. The transform faults have been discussed in detail by Allen[9]. The faults where the faulted medium is continuous and is conserved are known as transcurrent faults. The rift faults are mostly transform faults where without an increase in offset, the fault connects two expanding ridges.

If one investigates the magnetic anomalies of the ocean bed on both sides of a rift, a strip pattern emerges where strips of sea bottom seem to have been formed during alternate magnetic polarity epochs of the earth. This means that lava flowed on both sides of the strip in the same epoch but on account of the movement of the ocean bed, it is now positioned away from the rift clearing the way for the formation of the next strip. Similar in many respects to the ocean rupture rifts are the deep sea trenches. These too are deep narrow rifts going down about 5 km from the sea bottom. The trenches are only 5-10 km wide. The heat flow is weaker. Sometimes, there is continuation of the trench over land. The trenches are parallel to the island chains and lie on the convex side of the island arcs. The valley along the transform fault is the transform valley. The zone of interconnection which are presently active is called the principal transform displacement zone (PTDZ). At the intersection of the ridge with the fault, there is a deep depression called the nodal basin. The inactive portion of the fault is the fracture zone. Gravity anomalies are associated with the transform faults.

A subduction zone of the plate boundary is where the cold lithospheric plate is subducted into the mantle. It appears on the surface as deep oceanic trench or in a line of volcanoes on the overriding plate, e.g. the island arcs or the continental volcanoes to the west of the Andes. The depth of a trench may exceed 8 km as of the Mariana Trench. The oceanic trenches and island arcs are concave towards the overriding plate. The volcanic ring around the Pacific is due to the subduction of the Pacific plate.

8. Contiental Drift

Wegener[14] proposed in 1929 that the continents drift from the closely fitting shapes of West Africa and Brazil. By means of translations and rotations, most of the continents could be gathered into the shape of a single continent (Pangea) of the distant past. India was a part of the later day Gondwana land along with eastern Africa, Australia, Madagascar and Antarctica. Paleomagnetic pole reconstruction throws considerable light on the movement of continents.

6
Surface Magnetic Field of Mercury and Earth

1. Introduction

We have already introduced the subject of magnetism as well as magnetism linked to earth but not really dealt with the origin of the magnetic field on a planet. With the demise of geodynamo, its source of magnetism needs an investigation. The magnetic field of the earth has always attracted scientists, scholars and priests in various countries. The field was providing a navigation aid even when out of sight of coastal features. A cosmic magnetic field away from any coast has a strength about 5 nT. The field will be enhanced on land where it will become permeability times the cosmic value. Most matter is diamagnetic. There are, however, materials on land which have paramagnetic or ferromagnetic susceptibility and will correspondingly cause some magnetism to appear near them. We can calculate the expected increase of the field and begin with Mercury.

2. Magnetic Field of Mercury

Mercury is a solid planet. The field near its surface is 3.5 mG and the field away from the planet is 5nT. The enhancement near the planet can easily be explained by simple concentration by the low field permeability of the iron content of the planet. The iron core's radius is known to be 2/3 of the radius of the planet. The field will be 7 mG at the iron core (Parker[23]). Such an enhancement from the space field of 5nT can be explained by the concentration caused by the low field permeability of iron of 140.

3. Earth's Magnetic Field

The Earth has a surface field of the order of 1 G or the field at iron core would be more than 2.3 G requiring an increase due to a much higher permeability than the low field permeability of 140 satisfactory in case of Mercury. An additional field other than the space field has to be discovered operating in case of the earth. Nanda[24] had theorized in 2004, the existence of tritium atoms in the innermost core due to ternary fission of small proportion of the fissioning uranium that had caused the first atomic explosion. Tritium has a nuclear magnetic moment which gets aligned with the axis of rotation to produce the extra inducing field. The magnetic moment of the number of tritium atoms in contact with one square cm of the inner surface of the solid iron core multiplied by the magnitude of the Bohr magneton and divided by atomic diameter gives the pole strength σ and the extra field close to a plane surface is $4\pi\sigma$ which amounts to 5.3 mG. Using the permeability which is now for higher existing field than larger for the cosmic field. The field at the surface of solid iron core inside the earth is about 4G and the field is reduced at larger radius of the earth to about 1G.

7
Terrestrial Magnetism

1. Introduction

Any theory of terrestrial magnetism leads to the understanding of following characteristics: The axis of the equivalent dipole must be the same as of paleomagnetic field at that time which was inclined at 11.5° to the axis of rotation. The paleomagnetic field shows many reversals. At some random interval, the main dipole field decays to a low value and then reverses or recovers. Besides secular variations, the field shows certain anomalies that move westwards. About two billion years before present, the field became two-fold (Hale[2]). In certain areas on the surface of the earth, there are non-axial fields. The equivalent dipole is slightly displaced from the center of the earth. The theory should be able to give low value for surface field for Mercury, but a high value for earth. The popular dynamo theory of the last century has been over thrown not only because of viscosity considerations but also from the ancient field being inclined to the axis of rotation of the earth.

The earth had couple of atomic explosions in it. The explosions would also produce large amount of helium from the ternary fission of a fraction of the uranium atoms. The helium would be trapped near the center of the earth by the enormous solid iron shell. Still some of it is leaking out from across the entire earth, to the atmosphere from where it slowly escapes to space and eventually to the sun. The ratio R_a of ^3He: ^4He in the atmosphere is 1.38×10^{-5}. The ratio on mid-ocean ridge basalt is 1.17×10^{-5} (Kurtz and Kenkins[24]). Allowing for difference in diffusion rates, the ratio in the reservoir would be 1.02×10^{-5}. Helium stays as a gas since it cannot be liquefied by pressure alone. The density could be regarded as the specific volume per mole of 10 cc where the potential energy is minimum (Wilks and Betts[25]). In place of high compression, the pressure must be limited by distortion of the electronic shells in the helium atom. For the R value under discussion, 98,000 helium atoms surround each tritium atom and the inter-atomic distance between two light atoms would be 117 Å.

2. Duration of Stable Alignment

Only the collision with another ^3He atom can extinguish the spin alignment. Pair-wise collisions would occur. The magnetic attraction would cause a movement of the two atoms; the speed is given by dividing the force of attraction by $6\ \pi\eta a$ (Doi and Edwards[26]), where η is the viscosity and a the dimension of the atom. The diameter of the moving particle is 2.2×10^{-8} cm. The field at distance d along the axial direction from a dipole of moment M is $2M/d^3$. The force exerted by such a field on another magnet of moment M would be F = (Mdel) H, Kompanyet[27]. Thus for collinear dipoles, the force is $6M^2/d^4$. If the distance moved is x, the force is F = 6 $M^2/(L-2x)^4$ and the time taken by the atoms to collide in millions of years would be $t = (1/3.15) \times 10^{-13} \times$ the integral $\int_0^{L/2} 6M^a(L-2x)^b dx / v$

The magnetic anomalies show an interesting pattern on both sides of a rift. Ridges and rifts are associated with gravity anomalies and a deviation from isostasy. The unusual flow of heat from the rifts indicates ascent of magma masses deep below the oceanic crust. The Atlantic rise must have been formed by the ascent of magma along the rift that runs along the ridge axis. There are basalts present along the rise and the age of these basalts is only about ten million years. The rifts by themselves cannot accommodate the expansion of earth's surface without relative movement along transform faults. The transform faults have been discussed in detail by Allen[7]. The faults where the faulted medium is continuous and is conserved are known as transcurrent faults. The rift faults are mostly transform faults where without an increase in offset, the fault connects two expanding ridges.

If one investigates the magnetic anomalies of the ocean bed on both sides of a rift, a strip pattern emerges where strips of sea bottom seem to have been formed during alternate magnetic polarity epochs of the earth. This means that lava flowed on both sides of the strip in the same epoch but on account of the movement of the ocean bed, it is now positioned away from the rift clearing the way for the formation of the next strip. Similar in many respects to the ocean rupture rifts are the deep sea trenches. These too are deep narrow rifts going down about 5 km from the sea bottom. The trenches are only 5-10 km wide. The heat flow is weaker. Sometimes, there is continuation of the trench over land. The trenches are parallel to the island chains and lie on the convex side of the island arcs.

The valley along the transform fault is the transform valley. The zones of interconnection which are presently active are called the principal transform displacement zones (PTDZ). At the intersection of the ridge with the fault, there is a deep depression called the nodal basin. The inactive portion of the fault is the fracture zone. Gravity anomalies are associated with the transform faults.

A subduction zone of the plate boundary is where the cold lithospheric plate is subducted into the mantle. It appears on the surface as deep oceanic trench or in a line of volcanoes on the overriding plate, e.g. the island arcs of the continental volcanoes to the west of the Andes. The depth of a trench may exceed 8 km as of the Mariana Trench. The oceanic trenches and island arcs are concave towards the overriding plate. The volcanic ring around the Pacific is due to the subduction of the Pacific plate.

3. Reversals of Magnetic Field

The ratio R_a in the atmosphere of ^3He: ^4He is seen to be 1.38×10^{-6}. The ratio for the gases above mid-ocean ridge basalt is 1.17×10^{-5} (Kurtz *et al.*[28]). For obtaining the composition in the gas reservoir deep down, we have to divide the atmosphere figures by $(4/3)^{1/2}$ to account for the lower diffusion rate for the heavier isotope. Bosch and Huizenga[29] give the value $R = 1.5 \times 10^{-5}$ for helium given out in fission of uranium. For this proportion, 74,000 helium atoms (of mss 4) surround each light helium atom. Thus the distance L between two ^3He atoms will be on the average, the cube root of 74000 multiplied by the radius of the helium atom or L = 42 × 2, 53 = 106.3 Å. Only collision between ^3He atoms can extinguish the alignment of these. A magnetic dipole-dipole attraction exists causing a movement of such two atoms. The attraction would cause a movement of the two atoms, the speed is given by dividing the force of attraction by $6\pi\eta a$, where η is the viscosity and a the linear dimension of the atom ($a = 2.2 \times 10^{-10}$ m). The magnetic field H at a distance d along the axial direction from a dipole of moment M is $H = 2M/d^3$. The force exerted on another equal dipole of moment M is (Kompaniet[27]) is $F = 6\ M^2/d^4$. The two dipoles will collide when each has moved a distance L/2. When the dipoles move distance x, the force is $F = 6M^2/(L-2x)^4$. The force determines the speed v and the distance to be covered by the speed gives the time in second. Time T in millions of years will be given by converting time to million year units (1 million year = 3.15 × 10^{13} seconds). T will be

$$T = \frac{10}{3.15} \times \int_0^{L/2} \frac{dx}{v}$$

Using SI units the magnetic moment per tritium atom M = 2.1 nuclear magnetons where 1 nuclear magneton = 5.05 × 10^{-27}. The time t in million years is

$T = (2.2 \times 10^{-19} L^2/10)/[(3.15 \times 10^7 (2.1 \times 5.05 \times 10^{-27})^2]. \eta$, so $T = 2.7\eta$ million years. Since the average position of the interacting dipoles will be random, the numerical measure should come out greater than calculated. The viscosity of helium under high pressure may be considerably higher, these collisions will randomize the alignment and another long period will have to pass before the spins align again or be in reverse alignment. This can lead to a magnetic reversal or aborted reversal. Versub[30] has noticed in paleomagnetic measurement of magnetization of rocks, that time spent in excursion lasts few thousand years.

4. Non-Axial Field and Curie Point under Core Conditions

The field for which the generating dipole is non-axial had intrigued the dynamo theory. The portion of the magnetic field that does not reverse can easily be ascribed to some projections on the solid iron core whose tips are at a temperature below the Curie point on account of their dipping in cooler fluid that gives them ferromagnetic susceptibility and the reversing exciting field may not be above the required coercing field on account of the position of the iron tip on the hysterisis curve. The entire solid iron core must be just above the Curie point except for certain tips projecting into the cooler liquid.

Gutenberg showed in a diagram variation of density with depth with drops in density at the inner and the outer boundaries of the liquid iron core. There has so far been no experimental observation of Curie Point of γ-iron which is stable at high temperature, Theoretically, the fcc γ iron is suspected to have a Curie Temperature of 1800K which will rise with pressure. Leger[31] showed that bcc iron at a low temperature has temperature coefficient of the Curie Point of 1.5 K/k.bar. If one could make an extravagant extrapolation, this stable form might be occurring giving the Curie Point between 6400 and 6000 for core conditions so that the temperature of iron stays above the Curie Point.

5. Displacement of the Dipole from the Center of the Earth

It could be due to the non-central location of the helium reservoir after the massive atomic explosion in the innermost core. The centrally displaced exciting field will create paramagnetic moment in the core which will have cylindrical symmetry around the dipole in the helium reservoir and not around the axis of rotation.

The treatment of the magnetic field given above is in revision and amplification of the subject in the paper (Nanda[13]) published in 1977.

8
Fluctuations in Acoustic Signal as Guide for Deep Sea Fishing

1. Introduction

Deep sea fishing is carried out in known areas of sea from pure experience. So far there is no instrumentation for locating high sea fisheries. This problem was raised at a commonwealth conference on oceanography in 1954 at the British National Institute of Oceanography where I was sent to represent India. I had kept this need of fishing fleets of the world at the back of my mind till I had an occasion to be on the Downwind Oceanography Research Expedition of the California Institute of Technology. As an exploration effort, line fishing was carried out occasionally and the catch in limited time was recorded along with any observation of shoals and their extent, when they could be seen by the ship borne TV camera.

Fluctuations of sound transmitted in the sea had been a hot subject for the navy around the decade beginning in 1950. I had published the reason that the observed fluctuations were too large to be accounted for by any thermal micro-structure of the ocean. The only way fluctuations happened was because of the presence of varying amounts of air bubbles associated with fish or other structures and detritus connected with the biological content of the sea encountered by a pulse of sound in its passage from source to the detecting instrument. For sound transmission and recording, one could rely on the echo sounder pulses imagining the experiment was in relatively quiet sea and over a uniform bottom of the sea to give a standard echo. Thus it was obvious to me that the estimation of the biological content per unit volume should be connected to the variability of the echo amplitude. The echo sounder in our ship was fitted with automatic volume control. The engineers on board agreed to disconnect the avc feature so that fluctuations of the height of the echo could be measured. I have published the results (Nanda[32]).

2. Fluctuations of Bottom Echoes

The duration of the experiment at each station is so small, that conditions of sea are supposed to be the same in the few seconds of the experiment. The echo intensity is proportional to square of the recorded amplitude of on the pen recorder. When the observed variance is large, the spot must be right for fishing. The result was corroborated by actual fish hauls at each station. The method was noticed by the scientists but India had no deep sea fishing fleet. In actual worth, the stage was ripe to have some fishing vessels with the avc removed echo sounder. At the end of this long narrative, will restrict myself to underline two other important papers.

3. 9 Old Important Research Papers

(a) Origin of microseisms. (Nanda[33]) which also has an interesting discussion comparing it with the theory advanced by Longuet- Higgins. I have only to add that after we had both retired from the attachment to our respective theories, I had a chance meeting in USA with Longuet-Higgins and he on his own told me that long ago he gave up the mathematical derivation from interference of sea waves and conceded that since my ideas had the support of Prof. Gutenberg, the eminent explorer of microseisms, who favoured involvement of coastal rocks and winds besides the introduction of measurable sea surface roughness, Nanda's description of the origin is going to last.

(b) Evaluation of Crustal Characteristics like the average increase with depth of the P wave velocity (Nanda[34]). This was my first paper on a subject of geophysics which had assumed importance in early research work on crustal seismology a subject practically monopolized by DTM regarding the eastern sea board of USA around the years 1950-1952 when I was a

Carnegie Fellow there, Dr. M.A. Tuve and I often together occupied seismic stations to observe arrivals from shots fired by US Navy for crustal seismological studies of DTM, Dr. Tuve noticed the long tails of such seismograms and observed that there must be considerable scattering and re-bunching of waves from slightly different paths. I analyzed the amplitudes under various possible types of seismic waves. The decay of amplitude was linear in the tail of the record and the slope was indicating the slight average increase of velocity with depth. The result was hailed as a great advance since a single record close to the shot point could give the increase of velocity with depth eliminating the usual method of obtaining records at many stations at varying distances from the shot point. The ideas about various "phases" in the record had to be changed. Only first arrivals had a straightforward meaning.

(c) An article sent by Nanda[35] for publication in 1956 was published in 2012. An abstract retrieved by JGR and received from them recently is reproduced below as transmitted to me by J. Geophysical Research.

4. Seismic Exploration of the Continental Shelf Off West India

ABSTRACT. Some recent observations in connection with the seismic exploration of the continental shelf off the west coast of India are described. A single ship with a single sono-buoy was used. Five shots were fired at distances up to six miles from the sono-buoy. Roughly the depth of the ocean was about 80 fathoms and a negative temperature gradient existed. The exact times of the shots were calculated from the observed arrival of direct waves. The loose sediments have depth calculated from assigning a broad maximum in the records to a reflection from the bottom of the sea.

9
Satellites of Mars
A Test of the Theory of Planetary Atomic Explosions

The new idea of atomic explosions in planets is the central theme of my contribution to earth physics and a convenient test is offered by applying it to Mars which is fairly close to earth in size. We have seen that an explosion would explode out a scab of size S such that S = m R Θ/w where m is the mass of the planet, R its radius and w the angular speed of rotation implying that the satellite mass is proportional to the planet's mass. Dr. S.C. Solomon, director of DTM, has determined the masses of both tiny satellites correct to grams which means that the size of Mars must have been very small when the two satellites emerged. This is completely a different occurrence to what happened for earth. The difference is visible in the relative size of the moons of Mars and the Earth's moon. How did this happen? The answer is in the slight difference in the temperature of the corona delivered mass of earth and of Mars. The delivery is made by an assigned corona limb with possibly different temperature reached by the material carried by it. In the case of earth, the material is delivered at about the melting point of silicates close to 1000°C. We can infer that the material delivered to the site of Mars was at 100°C higher temperature which will deliver molten uranium that could quickly take its gravitation guided place at the center of Mars and Mars was soon ready to explode. For earth this took 730 million years as calculated in my paper on earth physics. Thus as soon as Mars was big enough it shed the two tiny satellites and had its two explosions which must have scattered the remaining molten uranium all over the small Mars at that time causing the surface of Mars to be like that of the earth's moon. This is how it has remained for billions of years and the observation in satellite pictures of water in canals or moisture on Mars are deceptive. The flows are of magma and we do not expect trace of liquid water even in the deep crevices on which future voyagers from the earth could rely with confidence. The volcanoes and meteorite impacts have dealt harshly with Mars. It began its history at high temperature that could remain high for at least a billion years, it being closer to the sun.

A confirmation is in the simple explanation now of the reputed intense bombardment by meteors or comets after it came up to full size. The much more simple explanation is that after 730 million years of the start of formation of the earth, there was an atomic explosion at the center of the fully grown earth and all the rocks and boulders that might have exploded out of the earth and did not go beyond the Roche limit. They fell back sooner or later on the surface of the earth and led to the insight that there was intense bombardment received by the earth after its full assembly.

References

1. J. N. NANDA (1997), *Geophysics* (ATextbook), pp 1-410 Ashish Varma Publications, Delhi.
2. C.J. Hale (1987) Archean Preterzoic Linear Core Nucleation. *Nature*, Vol. 329, p. 236.
3. Fred, S., Singer (1977) The History of Earth-Moon System. *Earth Sc. Rev.*, Vol. 13, pp. 171-189.
4. J.T. Wilson (1973), Mantle Plumes and Plate Motion. *Tectonophysics*, Vol. 19, pp. 149-161.
5. R.S. Dietz (1961) Continental and Ocean Basic Evolution by Spreading Sea Floor. *Nature*, Vol. 160, pp. 851-870.
6. Daniel M. Davis and Sean S. Solomon (1981) Variation in Plate Motion since Cretaceous. *Tectonophysics*. Vol.74, pp. 189-208.
7. V. V. Beloussov (1970) Against the Hypothesis of Sea Floor Spreading. *Tectonophysics*. 9. pp. 489-571.
8. Bhattacharya *et. al.* (1994) Evidence of Sea Floor Spreading in North Arabian Sea Earth and Planetary Sc, *Letters* 125, pp. 211-220.
9. T.D. Allen (1969), Review of Marine Magnetism. *Earth Science Review*, Vol. 15, pp. 217-254.
10. P.M. Hurley, *et. al.* (1962), *J. Geophysical Research*. Vol. 67, pp. 5315-5334.
11. G.V. Rao and R.U.M. Rao (1983) Heat Production and Flow in Basement Rocks of the Gondwana Basin. *Tectonophysics*. Vol. 91, pp. 105-117.
12. J.N. Nanda (1989), Earth's Innermost Core. *Proc. INSA*, Vol. 55, pp. 101-109.

13. J.N. Nanda (2011), *Origin of Terrestrial Magnetism and its Reversals*, Vol. 77, pp. 227-232.
14. A. Wegener (1929, reprinted 1966), *The Origin of Continents and Oceans*, Methuen, London.
15. K.A. Anderson, P.P. Lin, and R.P. McGuire. (1977), Linear Magnetic anomaly associated with Rima Sirsalis on Moon. *Earth and Planetary Sc. Letters,* Vol, 151, pp. 141-151.
16. L. Ian Lumb and K. D. Aldridge (1991), Viscosity for Earth's Fluid Outer Core and Core-Mantle coupling *Journal of Geomagnetism and Geo-electricity*, Vol. 42, pp. 93-119.
17. Sarvajit Singh and Ari BenMenahem (1963), Seismic Waves And Sources. *Springer*, Verlag, Berlin.
18. L.G. Holcomb. (1983), *5th Report*, IASPEI Commission on Microseisms.
19. S.K. Wilde *et.* al. (2001), *Nature.* Vol. 409, pp. 178-185.
20. P.N. Carla and Langmair (1988), Oldest Crust in Young Moon. *Earth and Planetary Sc.Letters.*, Vol. 40 and pp. 119-130.
21. R.A. Lyttleton (1984), *The Structure of Terrestrial Planets and the Tidal Friction Theory*. Reprints of University College, Cardiff. U.K.
22. M. Ishi and Dzievonsky (2002), The Innocent cor. of the East-Evikesee of change in Amisatopre Behaviour at the basis of c books P.N. A.S. Vol. 99, pp. 14026-14033.
23. David A. James (Ed.) *Encyclopedia of Solid Earth Geophysics.* Van Nostrand Co. p. 315.
24. M.D. Kurtz and W.J. Jenkins (1982), Helium Isotopic Systematics . *Nature*, 297, p. 43.
25. J. Wilks and D.S. Betts (1987), *Introduction to Liquid Helium.* Clarendon Press, Oxford.
26. M. Doi and S.F. Edwarda (1986), *Theory of Polymer Dynamics.* Chapter 3, Oxford.
27. A.S. Kompanyet (1961), *Theoretical Physics.* Foreign Language Publications. USSR.
28. L. Ian Lumb and K.D. Aldridge (1991), Voscont for Earth's fluid care. *J, Geomagnetism*, Vol. 43, pp. 93-110.
29. Robert Van den Bosch and John Huizenga (1973), *Nuclear Physics.* Academic Press. p. 379.

30. K.L. Versub (1982), Geomagnetic Excursions: *A Critical Assessment from Sediments of Brunhes Epoch*. Phil. Transactions of Royal Society of London, Vol. 306, pp. 21-35.
31. J.M. Leger, C. Loreta-Susse and B. Voder (1972), Pressure Effect on the Curie temperature of transition metals and alloys. *Phys. Rev.* Vol. B6, pp. 4250-4261.
32. J.N. Nanda (1978), Determination of Volume Scattering and possibly Biological content in the column of sea water below from sea bottom Echo Fluctuations. *J. Acoustical Society of America*. Vol. 63, pp. 618-620.
33. J.N. Nanda (1960), Origin of Microseisms. *J. Geophysical Research*. Vol. 65, pp. 1815-1820.
34. J.N. Nanda (1960), Evaluation of Average Crustal Characteristcs from Reverberation of Seismic Waves. *Proc. Royal Society of London*, Vol. 256 A, pp. 28-38.
35. J.N. Nanda (1960), Seismic Exploration of the Continental Shelf Off West India, *J. Geophysical Research.*

Index